Using Herbal Remedies

Page Left Blank Intentionally

Using Herbal Remedies

By: Spirit Walker

Dogwood Hollow Press
2018

Copyright © 2017, 2018 by Merlyn Seeley/Spirit Walker

All rights reserved. This book or any portion thereof may not be reproduced or used in any manner whatsoever without the express written permission of the publisher except for the use of brief quotations in a book review or scholarly journal.

First Printing: 2017

Second Printing: 2018

ISBN: 978-1-387-36036-9

Dogwood Hollow Press
Houston, Missouri 65483

www.amazon.com/MerlynSeeley/e/B008EEZ9QO

Dedication

I dedicate this book to James Wolf Seeley. Thank you for blessing the homestead with your little presence and smile that can light up a room. May all your days be healthy….naturally!

Contents

1. Acknowledgments v
2. Foreword vi
3. About The Author vii
4. Preface xii
5. Introduction 19
6. Chapter 1: Four popular issues and how to treat them 25
7. Chapter 2: How to make your own herbal medicines 53
8. Chapter 3: Herbal replacement for 5 popular OTC drugs 65
9. Chapter 4: Herbal treatments for children 87
10. Appendix 1 110
11. Reader Notes 113

1. Acknowledgments

I would like to thank all my friends that are herbalists and have shared their knowledge with me to improve my own throughout the years. Through community sharing we can blanket the world with natural healing and great health. May all your days be filled with flowers, herbs and long lasting natural health!

2. Foreword

3. About The Author

Merlyn Seeley AKA: Spirit Walker has been writing about how he lives and living what he writes for more than 7 years, off the grid, in the southern Missouri Ozark forest with his family. He writes an off grid column for his local newspaper since 2017. He and his family have went back to the land. Spirit says,

"Back to my roots, and back to when life made more since, had more meaning, was less hectic, less stressful and way more healthy".

His writing is a way of practicing a self sufficient way of life as well as informing others of how they, too, can live a simpler way of life. A life more sustainable, self sufficient and meaningful. He is a published author, with over 40 kindle books with 6 in print, to date. His books can be found online, where ever books are sold, and are both enlightening and awakening. He is also a renowned Herbalist, Cherokee medicine man, Medical Specialist and natural health guru and uses nature to heal as well as what medicine the land can provide. He can trace his heritage back to his Cherokee and Black Foot Native American ancestors and so keeps the old ways alive at the homestead, in many aspects.

He enjoys researching online, to further his self sufficient knowledge and spending time on his homestead growing organic/heirloom food and herbs to be made into medicine, raising animals, wild foraging for more food and medicine and working directly with nature as a steward of the land. He has written for many high profile places online, as a journalist, alternative media writer and freelance writer for many years. Places such as Examiner.com, CBS news local, Yahoo! Contributors, Truth News, Natures News and worked as a natural living and herbal medicine expert at liveperson.com.

He has had numerous pieces published in

high profile magazines such as American Survival Guide, Backwoodsman, & Self Reliance magazine. Spirit Walker has an uncanny since of survival and has had many years of survival training in both the USA and USAF working in the military intelligence field doing secret and top secret operations state-side and (worldwide through computer operations), in his younger years. Now in his prime he has devoted his life to writing, living off grid, growing/raising his family's food and medicine, raising his daughter and grandson and showing others just how they, too, can obtain true freedom and a since of satisfaction with life and living in general.

To contact Spirit Walker via snail mail or to purchase additional signed copies or signed copies of his other books in print, send your letter or your request with payment of $15 (**cash only, any other form will be returned**) per <u>signed</u> book to:

Spirit Walker

6970 Lundy Rd.

Houston, MO. 65483.

You can follow him online at :

Facebook (Dogwood Hollow Homestead)

Word press (dogwoodhollowhomestead.wordpress.com)

Twitter (Dogwood_Hollow)

You tube (Dogwood Hollow Homestead Off Grid)

4. Preface

Growing up in the 20th century has to be something that those in the past would scarf at. What, with all our made up diseases, disorders and the drugs we have made to pretend to treat those made up things. There was a time in our civilization's history where people were healthy, stronger, had more stamina, energy and endurance.

It wasn't to terribly long ago either. There was a time when treating illness took a simple walk either into the forest or into the family garden for some herbs and foods that had proven their power over the course of 1000's of years. People were sick for a shorter duration and less often. Kids did not mature physically so fast either.

People feed their children meats they buy from the local big corporation grocery store loaded with hormones and steroids and the children grow breasts while they are still little girls and mustaches when they are still little boys. This is not natural. It is the result of a society that has become dependent on chemicals. Then those same people complain about those issues, and have no idea what the truth is.

We eat antibiotic laden meats from the grocery store all the time, as the FDA and USDA allows higher and higher amounts of chemicals to be sprayed! And then they wonder why their medical antibiotics are not working on today's infections! It is because they have overused antibiotics to a point that our viruses are becoming super bugs, and those same folks have no idea what the truth is.

So what is the answer then? It's quite simple really, natural remedies. Not only have these remedies been around longer than anyone that will ever read this book, but they are still as powerful today as they have ever been. There are no side affects to most natural remedies so you don't have to take a pill to counteract a pill you've taken.

There is no dependence to worry about with natural medicine, so use all the **Arnica Montana** you need for pain and don't worry. The natural antibiotics work on the super bugs they have created, from over using antibiotics, too. Simple but powerful plants grow in the wild and can be grown in your yard that are just as powerful as most our chemical antibiotics today.

Chronic issues can be treated successfully with natural remedies. Modern day medicine, that is usually prescribed for this type of issue, does nothing for the root of the problem. So the chronic problem never goes away. Why do you think they call it chronic pain or chronic illness? Because it is not meant to go away. They want you to believe that you have to live in pain or with the problem all of your life.

They tell you to just take their medicine to treat it, which is a lie. Modern day medicine treats the symptoms, NOT the root of the problem. For instance, most all pain is associated with inflammation in the body. But do they treat you for inflammation? NO, they ask you what your complaint is and when you tell them you have lower back pain they give you PAIN medicine.

What should they give you? For starters, something to counteract the inflammation that is causing the pain. To be honest here, if they did that, how much money would they lose off patients with chronic pain? Do you know how much money the pharmaceutical industry makes a year? It's a multi-billion dollar a year business! A wise man once said, *"Good health is great, but it doesn't make a lot of money"*.

If you are in denial about all this, then you can be sure you are a skeptic and that is just how they want you to remain. They do not want people knowing that the same little 2 foot tall purple cone flowers they call wildflowers and spread all along the sides of the roads, in your city, is Echinacea. Echinacea is a very powerful antibiotic herb! It even has the power to cure pneumonia.

I wrote *"Using Herbal Remedies"* because I am a natural healer, a self taught Herbalist and Native Medicine Man. It is my sole purpose on the earth to heal people. I teach others how to heal as well. Hopefully this book will do that for you. People come from all over to learn how to heal naturally and to get healed naturally. More and more people are waking up and learning the truth about modern medicine.

Yes, modern medicine has it's place, in saving lives. BUT when it is 100% clear they have no idea how to treat simple illnesses and how to prevent them by making sure the immune system is strong through nutrition, then there is a problem.

And as a final note, for all those that would scarf at my words, *"who does this guy think he is"*. Well I am a natural living specialist, I have healed countless lives with not only herbal remedies but using energies, food and meditation as well. I was a medical specialist in the United States Army for 15 years and so I have been on both sides of the fence.

I live off the grid with my family in the southern Missouri Ozarks and I grow, forage and produce 100% of my family's medicine. My homestead has been a beacon of light for those wanting to heal, learn to heal others and for natural medicinal supplies/herbs for many years. All I wish for is for all mankind to truly heal, and to do that you have to get to the root of the problem. The best part about herbal remedies is they treat the problem not the symptoms!

-Spirit Walker

Spirit Walker

5. Introduction

What is it? Ginseng plant

Congratulations on picking the book *"Using Herbal Remedies"* as your next book to read. In this book I will walk you through some of the most needed herbal remedies showing you what you can use to take care of some pretty popular medical issues. From a common headache to a common stomach ache I will tell you what herbs to get, how to prepare them and what you can expect from them.

In chapter 1 you will learn about various herbal remedies, pair up your issue with those in the chapter and see what it is that you can do for yourself naturally. In today's world it is very important not to get all caught up in all the false teachings such as, *"you have to use synthetic, man made drugs to deal with your medical issues."*

God put the plants on the Earth for us to use. Herbs are abundant on the Earth and if we all knew how to heal ourselves using wild herbs we would all be healthier. Although making herbal medicine should be left to those that have studied and learned the correct way to do so, most anyone can learn to make their own herbal medicines to be used for remedies for common ailments.

So here is the book, *"Using herbal remedies"*. I will also walk you through the steps involved in making your own medicines. Whether it is herbal tinctures, salves or infusions that you seek knowledge in, it is talked about in this book. Why take all those life altering pharmaceutical drugs, that only serve to make you sicker, when you can go right out to your garden or lawn, forest and get what you need, <u>for free</u>, and usually with out any side effects?

Of course you need to understand that it is possible to be allergic. If you are allergic to the ragweed family, then probably Chamomile is not an herb for you. BUT the beauty of that is, if you can not use one herb for some reason of another, there is another herb that can also be used, in it's place, that can produce the same affect.

Herbs may not be the answer to everything, but for the most part there is nothing that can not be treated using them and there are multiple herbs that can be used for the same reason. So if one don't work for someone, another most likely will. Diseases that supposedly do not have a drug to cure it probably have an herbal remedy that will.

In fact some people have learned about the long known about herbal cancer treatment called **Essiac Tea**. Although there are natural alternatives to the drugs we all know about, not many people usually choose to use them. Today all that is changing and more and more people are using herbal remedies than ever before. People are tired of remaining ill even when they are using a synthetic drug. And they are tired of taking a drug just to counteract the side effects of another drug they take.

Why would anyone want to take a drug that is going to make you sicker and then take another drug to remedy the side effects of the first drug and then deal with side effects of that drug? In, *"Using herbal remedies"*, I will tell you what you can do and how to do it instead of becoming their next lab rat for the next new, made up drug created for a made up medical issue/disorder.

If you want to know what you can do naturally for depression, insomnia, or infections, pain etc. This book will help you. It is not meant to be an herbal Bible, per say. Instead it is meant to guide you along the natural healing path and teach you how to

Spirit Walker

make your own natural medicine and how to use it properly.

6. Chapter 1: Four popular issues and how to treat them

What is it? Peppermint

Digestive Problems

When our stomachs are not happy, we are not happy. There are a number of common ailments that one could deal with when it comes to the digestive system. Did you know that more people take a drug for an upset stomach than any other digestive problems? Did you also know that almost every over the counter (OTC) drug for digestive issues contain dangerous heavy metals, and most of them do not work all that well? That is because we are not made to take synthetic drugs for problems with our digestive system, we are made to use mint family herbs!

That's right if you have an upset stomach, heartburn, bloating, gas, or any other problem in your stomach, intestines, or any organ associated with the digestive system, then you need either peppermint, spearmint, wintergreen, cat mint, horse mint, ginger, basil or one of the many other herbs in the mint family.

Although there are many ways in which you can take an herb that is part of the mint family and use it any way you want, the best way to get the mint's healing power to your

What is it? Spearmint

stomach quick is to drink it. So an herbal infusion is the answer here. Besides it has been proven that just drinking a warm cup of liquid can help calm an upset stomach. So paired with the healing power of some mint your stomach will thank you.

Now if you are dealing with nausea, such as air or car sickness, you can use ginger or basil to calm that right on down. In fact did you know that when you are on a plane and are getting air sick, the stewards will give you a ginger pill if you were to ask for something for air sickness? Next time, be ahead of the moment and carry some peppermints and ginger Altoids with you and pop them as needed.

Just be sure they too are natural as many OCT remedies, like those, are bad news and can only cause more stomach issues in the long run. To really help a child who is vomiting, get the child to chew on some fresh basil or oregano leaves or simple make a tea from those herbs. To make the basil or oregano tea a bit more palatable, as those are bitter herbs, also add in some peppermint leaves. Children love the taste of peppermint tea!

The vomiting will be curved a bit and they will feel better too. Stomach aches of all types, Irritable Bowl Syndrome, which is a reaction to eating genetically modified foods (GMOs), can be beaten using bitter herbs and those of the mint family. The mint family may seem small, but in actuality it is a very large family and there are more mint herbs than you may think. It is good to learn to use more than one.

What is it? Echinacea

Infection problems

A lot of people deal with different forms of internal infections every year and a large percentage of them go to the hospital and tell their doctor to give them something to help them get better. Then they dish out a ton of money for the doctor visit, the medicines he prescribed them, and then they go and fill the prescription and start popping harmful pills. Yea the infection goes away, in due time, but what about all the side effects that come with it?

What about the long term effects on the body and how over use of antibiotics have caused an influx of super bugs that we can not combat. Most people have an herbal remedy for almost every internal infection in your home right now. Garlic, alone, acts on most all internal infections like a Navy destroyer sent out to seek, find, and destroy.

Eating cloves of garlic may not be such an easy thing to do, but using chopped garlic in your foods and taking garlic supplements are a good way to get the benefits of the garlic. However, for those that have no problem eating raw garlic cloves, it is a sure way to kill that infection! For everyone else, I highly recommend you try garlic oil capsules the next time you find your self with something like bronchitis or pneumonia.

What is it? Grapefruit Seed Extract (GSE)

Another herb you can use in the form of a tincture or tea is Echinacea, it's a natural antibiotic and immune system booster. But the one that comes recommended the highest is GSE or Grapefruit Seed Extract. The little drops that you put in a cup of hot tea or some cold fruit juice is so concentrated that all you usually need for an adult dose is 15 drops.

Use this three times a day and in 3 days you will really notice it has been working, in as little as 5 days you will usually be infection free! I can not count the amount of times I have proven this in my own family and self. For a child dose, simple lower the dose to 10 drops, three times a day and for a toddler or small child, something like 5 drops, three times a day will do the job. The name of the company that I trust most is called Nutribiotic, the name alone hints at it's use.

GSE also works for infections outside the body as well. As a salve, essential oil, or added to a natural oil such as Olive oil, as a carrier, or even as a wash, you can effectively treat an external infection. When you use Echinacea or garlic externally, it's as simple as making a tea, allowing it to cool and then applying that directly to the affected area.

If you drop the essential oil of these plants into a carrier oil you are using it as a salve which will allow you to make the remedy stay on something, such as a wound, for longer periods. Try it with a bandage!

What is it? Valerian Root

Sleep problems

Everyone likes to sleep well. It's natural to fall asleep, stay asleep and be able to get a good nights rest. It is not natural that you have problems sleeping but so many people complain about the quality of their sleep. No matter what is on your mind, if you do not calm your mind and relax your body before laying down to sleep then you will not get good rest. There are many natural ways to get a better nights sleep. There are also natural herbs that can help your sleep problems as well.

One of the best herbs to use is called Valerian Root. Now Valerian Root is a mild sedative, put in laymens terms, it will help you to fall asleep, stay asleep and to sleep deeper and wake up feeling rested. Valerian Root is solely a medicinal plant and can be dangerous if you are not careful with it. Like other natural sedatives you do not want to be taking any sleeping pills or any sort of depressants to include alcohol when you are using Valerian Root.

If you use to much of this medicinal herb you can overdose and become quite ill with it. It is best taken as either a capsule, in which you would want to follow the directions on the bottle for use, or taken like a tea which is really an herbal infusion.

When it is drank as a tea you will want to drink it warm and about 30 minutes before bed time. I suggest taking about 8-16 ounces of this tea for sleep issues. Since Valerian Root can be quite harsh to drink alone, you may want to mix your Valerian Root with some other herb that balances out the harsh taste. Peppermint seems to be a very good herb to use with Valerian Root. Valerian Root comes in a tea, a tincture, capsules, and a bulk herb.

If you are into growing your own herbs, you can get Valerian plants as well as seed for a pretty decent price.

If you are not into Valerian Root, then another, much easier to drink herbal infusion that works with most folks is Jasmine tea.

What is it? Jasmine Tea

Jasmine is a flower that grows abundant. Although you can not grow just Jasmine tea itself, it is possible for you to make the mix yourself. But it is easier to simply buy Jasmine tea and use it as needed. For those of you that are interested, a cool little piece of information on Jasmine tea is how it is made.

Jasmine tea is actually black tea infused with the scent and oils of Jasmine flowers. Tea makers pick the Jasmine flowers after the sun has went down because the flowers only bloom at night. Then the flowers are tumbled with loose black tea many times to make the infused tea. Then the black tea is sold as Jasmine tea. The oil in Jasmine is simply amazing stuff. It acts as a mild sedative and calming herb.

Those who use Jasmine tea to get a better nights sleep swear by it. Drink a cup of the tea about 30 min. before bedtime and then go to sleep as normal. Jasmine is so mild it can be used during the day to calm yourself or a child, relax anytime and before meditations to aid you in relaxing your mind. It is not a good idea to drink Jasmine tea before driving or operating dangerous machines.

Jasmine can be found in most all herb shops, natural food stores, as a tea in most stores, online and in tinctures and essential oils. Jasmine tea is safe enough to be used all the time and for children and toddlers alike. Watered down, it can even be used to help with your baby's colic and to help baby sleep better.

What is it? St. Johns Wort

Depression/Anxiety

No one wants to be depressed, sad or anxious. These feelings make use feel bad and we just do not know what to do about it. Depression and anxiety can ruin your life if you allow it to. It can become chronic and a big problem over time. When you are feeling the blues you may want to do something but do not want to deal with using any dangerous drugs. There are many natural remedies that can help you fight the blues, but above all, the greatest remedy in my mind still has to be the herbal remedy of St. John's Wort.

St. Johns Wort has been coined in certain countries, such as Germany, as the miracle herb for depression. In Germany it works so well that the country endorses it and has allowed it to be prescribed by doctors for depressed patients. St. Johns Wort grows wild in moist woods just about all over the country. But you can get St. Johns Wort to use, yourself, from most all herb and natural food stores. It is plentiful and if you find the plant in the wild, take it as a blessing and do not over harvest it.

You want it to be there when you really need it. Using St. Johns Wort tea, tincture or extracts you can successfully cure your depression and make sure that it never returns. Taking St. Johns Wort capsules/extract every day when you are feeling down and out can pick you up in a matter of 15 minutes. As with any herbal remedy if you are pregnant or nursing a baby be very careful with how much you are using. Personally, I have used this remedy with 100's of private clients and know of it's power.

What is it? Kava Kava

Another awesome herb for serious anxiety control is called, Kava Kava, or sometimes just Kava.

If you are really looking for a way to relax and take the edge off of a headache or some other type of nerve pain, Kava is an all natural herbal remedy that an herbalist might prescribe you. Kava is a plant that grows in the Amazon Rain Forest and so Kava is not so easy to find here in the US. Another reason it seems to be so hard to find is because it works so well most herb and natural food stores are always just sold out of it. Once it is sold out it is hard to get replaced, but over time it will be back in stock. It is just best to get your Kava online. Now as an Herbalist, I will tell you that the best way to use this herb is to get it in the form of a pharmaceutical grade tincture. Although there are Kava extracts and teas and you can get the herb in bulk, the tincture will be blended perfectly.

Using Kava a person should be careful. Some studies suggest that Kava likes to tax the liver, meaning that it is hard on the liver because it contains what is called Kavalactones which are slightly toxic and more so at high doses. Although Kava is safe to use for anxiety and stress and even as a sleep aid and nerve pain remedy. More and more people are learning about the health benefits of this wonderful medicinal herb.

Users of this herb as a tincture should try it in a warm herb tea that has an easily palatable flavor. Kava will numb your tongue and make the inside of your mouth feel warm and less sensitive but the side effects are harmless and last only a short time. As such, this is also a great herb to use when making your own cough syrup or drops. Still the herb's power on stress and anxiety can not be matched.

What is it? Passion Flower

Maybe you are into more exotic ways of dealing with your anxiety and depression, but you still want the healing power of a natural plant. Passion Flower might be a beautiful flower with some really great tasting fruit, but did you know that it is also a very potent medicine for those suffering from anxiety and depression issues? It's true and in many places this flower even grows wild and can be harvested at the end of summer for use.

And if you are dealing with stress and anxiety and are looking for an herb that can be used anytime, by anyone, with no side effects then you might want to check into getting some passion flowers. They are available as a dried herb in host herb shops and online and are best when mixed with some other favorite herbs and infused into a warm tea.

When you drink passion flower there is a slight grape like flavor you will notice and then you will see that the flowers have the ability to calm and relax you. There are indeed many herbal teas on the market that contain passion flower. As I mentioned in the beginning, it also grows wild in most places around the country. Passion Flower's scent and flavor will do well with children of who's parents are looking for an alternative to calming drugs, due to it's grape like flavor.

Although I have never seen just passion flower herb tea, it could exist. But the best way to get some use out of passion flower is to get it mixed in with some other herb tea. Since you can buy passion flowers on the market in bulk then you can mix your own stress relieving/calming tea and drink your way to an herbal get-away.

What is it? Hops

As with all herbs there are multiple ones which can do the very same thing, as well as fix multiple issues. However, as I have probably mentioned before, not every herb works the same for every person, so you might have to find what works for you. But there is one that is right for you, for sure. Maybe your best choice would be a popular herb that is used in making one of the world's most popular beverages, beer. Yes I am talking about Hops. You can buy hops all by themselves too, and they can be made to be used as medicine for anxiety AND as a sleep aid, although I personally would not recommend this herb for those suffering from depression, because to me, it seems it could make it worse. It IS added to the world's most popular depressant…beer.

And when you think of hops most people think of beer. It is the very same ingredient. Now as a beer additive, there are other reasons besides just that, but you always were taught in school that alcohol is a depressant. When you are feeling like you need to relax and "chill out", you could get some relief from a cup of tea that includes some hops herb. Hops is a small white, tightly grown cluster that is dried and sold as an herb for stress, anxiety, and as a sleep aid. You can find some hops in a lot of herb and natural food stores and, of course, online.

There are tinctures and extracts, as well as, tea made of hops. Although I have never heard of just Hops Tea, you can buy quite a bit of herbal teas that include hops as an additive for stress and anxiety and a gentle sleep aid.

7. Chapter 2: How to make your own herbal medicines

What is it? Warm herb tea (Infusion)

If you are going to be successful at healing you and your family's issues using all natural means, in this case, herbs, you have to know what you are doing. Wrong doses and preparations could result in wasted herb and some medicinal herbs are pretty expensive. You could render the herb unusable and it's potency be worth nothing, too. So it is imperative that you practice and get to know what you are doing and with what herb you should be using for what ailment.

So we will start with how to properly brew the herbs to remove the medicinal qualities each one has. Some herbs come as hard woody parts, such as Valerian Root. Being a root means it is harder to remove the medicine, so there are specific ways. Some herbs come as soft flowers, those types it is easier to get the medicinal benefits, so a different preparation is needed and so on.

Decoctions

A decoction is sort of like a tea infusion but with a bit more work involved. The flavor is more intense and the medicinal content is a lot higher. The difference between an infusion and a decoction is in how it is made as well as what part of the plant you use. Decoctions are made from the woody, hard parts of an herb. And in some cases, a decoction is made using the berries of a plant. The reason you would use this method, with these parts, is because in order to get the medicine out of the plants parts you need to boil them. This is just what a decoction is, boiled herb parts. Once you place the hard parts, such as: bark, berries, or roots into a pot and add water you would boil the mixture for approximately 15-30 minutes.

Once you boil the herb parts the medicinal content is extracted and it will be very hot. Allow the mixture to sit covered as it cools. Make sure it is covered because the medicinal content can leave along with the steam. Once it is cool enough for safe handling you can strain the hard parts of the herb from the medicinal liquid. Now you use the liquid just like you would any herbal tea. It can be drank as well as used to make some of your other herbal medicines. Sometimes a decoction has to be taken by spoon fulls every few hours, so research and learn how to use the specific herb you are dealing with for the specific ailment you are treating.

What is it? Herbal infusion

Infusions

Infusions are herbal teas, literally. When you have a plant that you are going to use for herbal medicine or to make any sort of herb beverage you want to infuse the soft parts. Soft parts of an herb include some berries, the leaves and flowers. These parts are soft enough that they will give up their medicinal content quite easy. To make an herbal infusion all you would need to do is place some soft herbal parts into a cup and add boiling water. Allow the infusion to sit up to 5 minutes for herb teas and up to 10 minutes for medicinal herb teas. Always cover medicinal infusions as they are brewing to keep from losing the medicine in the steam.

Some herbs to use in your herbal infusions would be St. Johns Wort, Jasmine Flowers, Bergamont, Horse Mint, Peppermint, Spearmint, Camomile, Passion Flower, etc. You can also combine an infusion with a decoction if you have two separate herbs that you have to use and you have hard parts from one plant and soft parts from another. What you would do in this case is make your decoction and strain it. Once it is done you would use the hot decoction as the liquid to infuse your soft parts. Strain out the herbs and you have yourself an herbal infusion.

What is it? Herbal tinctures

Tinctures

What is a tincture? A tincture can be made using any part from the herb. Although some people will say that is is best when made using hard parts of the plant, you can make an herbal tincture from soft parts as well. The idea of a tincture is to extract the herbal medicine from your herbs at a concentrated rate so that you get a high dose of the medicine. Tinctures can be made using a variety of liquids such as water, alcohol and oils. My favorite is the alcohol tincture. Using something such as, 80 proof Vodka, you add the herbs that you want to use and seal it up. The herbs will sit in the Vodka and brew. The alcohol will extract the medicine from the herbs over time. In order to make an herbal tincture, of something such as Goldenseal, you would use about 2 cups of dried herb to about a quart of 80 proof or better Vodka.

Seal the top and allow it to sit for 2-4 weeks to steep. While it sits, make sure to have it in a cool dry place and make sure that once a day, every day for 2-4 weeks, you shake the bottle. This will make sure that the herb stays mixed up in the liquid because it will settle to the bottom. Once your tincture has sat for the desired amount of time you can strain out all the herb and place the tincture in your medicine cabinet. You can also place some in a medicine bottle with a dropper with the rest stored in the jar for later use and refills. A good dose for an alcohol tincture is about 4 dropper fulls or about 4 teaspoons in a cup of tea. You can use a tincture in a cup of tea or a cup of juice.

What is it? Herbal salve

Salves/Oils

An herbal salve/oil is a medicine that can be used on wounds, skin infections, cuts, scrapes and the likes. Basically a salve/oil is a tincture mixed with a "vehicle" to get the medicine to your skin wound. Tinctures alone would be ok except as a liquid it would just run off the wounds. As a salve or oil the medicine will be applied to the skin and the salve or oil serves as a holding agent for the medicine. This way the medicine will not just run off the skin it will sit on the wound and be absorbed by the body. After you make your tinctures, or if you are going to be using an herbal oil, you will need something such as beeswax to create the medicinal salve. If making an oil, a good carrier is olive oil or coconut oil.

For the salve, simply heat the beeswax up, add in the desired amount of tincture or oil and allow it to cool in a small container that will hold the mix. Once it cools it will be easy to use a cotton swab, your finger or a tissue to apply the medicinal salve to the affected area.

If you are making an oil, there are two ways. My favorite is to take a tincture and add drops into the oil. I store the infused oil in a jar with a nice fitting lid and use it when I need to. Another way is to take a jar, fill it half way with dried herb and then the rest of the way with either olive or coconut oil. Allow the oil to sit, in this case, for a month. It will infuse and then you can strain the herb out and put your medicine in a jar with an air tight lid to be used when needed.

8. Chapter 3: Herbal replacement for 5 popular OTC drugs

What is it? Home apothecary

Over the counter drugs or OTC drugs, are a massive money maker for the pharmaceutical industry. Rightfully so, because they are so readily available and anyone, of legal age, can purchase them anytime, for any reason. They come with their own bleak, at best, instructions for taking them properly and with a small insert that is hardly ever read. On that insert it tells you of the side affects associated with the drug. There are warnings, rules for taking this dangerous chemical laden drug that is supposed to miraculously cure you of everything from a headache to a stomach ache and more.

This is all done using the drugs chemical make up. Because it is unnatural, made in a laboratory as opposed to grown in the forest or someones organic garden. You are told to take the drugs and your problem will go away, such as a headache. I ask, how does a drug bottle know why YOU have a headache? Last I checked, OTC drugs do not come with their own cat scan or miniature doctor that looks you over and locates the problem. If that were true, half of the patients that would have taken the OTC drug, would be told to take a different drug. I say that because OTC drugs are made to mask your problems, not fix them.

What is it? Societies idea of free thinking

If there were a miniature doctor with each bottle on the shelves, he could look at each individual person and possibly tell why YOU have this headache, which could be linked to something simple or worse, depending on who you are. What I am trying to say is every bottle of OTC headache medicine can not and is not for every single headache that walks through the store's doors. But everyone buys this drug as if that is true, and you trust it. People are told to trust it. You are brain washed into believing this lie, that a bottle of Tylenol, for example, will cure all your headaches, and it does!

Where is the lie you ask? Simple, your headache is not "cured" it is covered up. So it is easy to say that little bottle is for everyone with a headache, because no one knows why they have an achy head. So its much easier to just cover up the problem, let it heal itself, which most headaches will do over time, and then move on. It's a way they can sell you OTC drugs, successfully. Don't you see, it's a gimic. They can let everyone that comes in with the same complaint just choose what OTC drug they want because it is not going to be tailored to anyone's problem, it is simply made to do the same for everyone that buys it…cover up the symptoms.

Something as simple, usually, as a headache is mostly linked to something in your environment that causes you the pain. The pharmaceutical companies know this and know that in time the headache will clear up on it's own. So letting you take some OTC, such as Aspirin, won't hurt them…but it will you, in the long run. Since they are not really curing you, they are not hurting their repeat business. You WILL be back because you are not healing your problem, you are covering your symptoms. Same thing doctors do when you go see them and they write you a prescription. They do not cure you because that would mean no more sick people, they would be out of a job. Modern day medicine, doctors, hospitals all they are made to do and taught to do is mask symptoms so the repeat business is always there.

You want to heal! Herbs do that. They get to the root of the problem, if you take the right herb for the right reason, and fix what the issue is. So no more delays, here are four common OTC drugs and their herbal replacements. Keep in mind, these herbs do what the OTC drug does, mask the problem. If you want to get totally healed, you have to first find out what the root problem is and then use the proper herb to fix it. However, for those that want to use all natural drugs, here is the list:

What is it? White Willow Bark (natural aspirin)

Tylenol/Aspirin- White Willow Bark. This is actually natural aspirin. It is where the original aspirin came from and still today is a major player in the aspirin arena. White willow bark is usually found in bulk at natural food stores, herb stores or online and home users will mix the bark with 80-100 proof alcohol and create a tincture. Still some users will just boil the bark for 10-15 minutes, covered and allow it to sit, covered, for 15 minutes before adding a bit of sweetener and drinking it.

White Willow Bark is natural aspirin, plain and simple and it will alleviate most headaches, body aches and pains and a host of other issues. It is worth in depth research on your part. Find yourself a willow tree and take the inner bark, that is the right part. In the good ole days, people would just take the bark that way and chew it, as well.

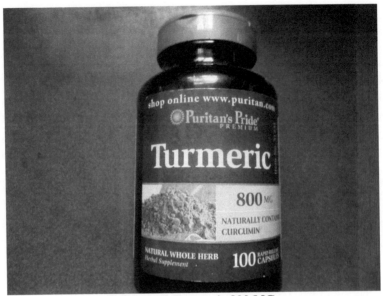

What is it? Turmeric 800 MG.

Motrin/Ibuprofen- Turmeric. Ibuprofen works because it actually takes down your inflammation. Inflammation is associated with almost ALL pain in your body. So to get rid of pain you have to get rid of inflammation. If you knew that the last time you went to the store and grabbed a bottle of Motrin 800, you could have skipped the massive tax on your liver that stuff causes and sprung for a cheaper bottle of pharmaceutical grade turmeric instead. A food that has the ability to heal your inflammation, in high doses. This is why curry dishes are so good for you. It is why, in the country of India and other places where curry dishes are the main source of food, that you don't see a lot of inflammation issues, not a lot of pain problems in India.

Seriously, grab a bottle of 800 mg turmeric and take the stuff for a week, see what happens to your body. You will discover almost all your pains just go away.

What is it? Boneset herb

Aleve/Naproxen- Boneset. This is a pretty strong medication with some pretty strong side affects over short periods of use. Have you ever read the insert that comes with this stuff!? It's incredible how detrimental this stuff is on your body organs. Why do you take this stuff? Bad headaches, migraines, back aches, knee pain. You name it, people use Naproxen type drugs for on going pain and to have pain relief all day. Yea it gets rid of the pain, and yea it works a long time, probably not as long as they claim, but close to it. Problem is, it's just a numbing agent. Migraine headaches are a sign, a statement, a warning to you from your body that something is not right. If you ignore this by simply covering up the symptoms, the root cause will not only still be there, but could be getting worse with each bottle you buy.

Instead, try a natural product that has proven to be effective. At least if you are going to cover the problem up, do so naturally. What you need to do is find the problem, then using a natural remedy, fix the problem and quit buying Aleve and Naproxen to mask your symptoms. Arthritis is a major problem in the US. Aleve promises to give lasting relief. But did you know arthritis is usually inflammation that has been left to go nuts? There are foods high in bromelein, a natural occurring chemical, that can fix your arthritis over time. There are herbs that can do the same thing. There are even lifestyle changes that do this and exercises too.

But if you need relief now, use one of the replacement herbs. They are pain relief herbs that will work immediately. The side effects won't be as bad as the OTC drugs and you can take the herbal replacements as you need too without worry of overdose or attachment. The herbal replacements are cheaper, work better, and won't support the pharmaceutical industry.

What is it? Chamomile

Hydrocortizone creams- Chamomile cream. Plantain cream. Itchy skin is something that millions of people can say they have experienced. Children, adults, it does not matter, at some point in your life you are going to have an itch that causes you discomfort and eventually you will want to control that itch. I once used chamomile cream, that I made myself on a baby who's parents did not know what to do. They had been to the baby's doctor and the doctor had prescribed steroids for the baby. They had tried all the OTC creams, which usually contains steroids as well, or aluminum, a heavy metal that soaks into your skin and poisons your blood.

I mentioned letting me give the child a bath in chamomile tea. Just water infused strongly with chamomile herb, was all it was. I washed the child, who was suffering with psoriasis and that night the child slept. I was told that the child had not slept through the night in a long time because of the itch. In an earlier chapter of this book I taught how to make herbal salves and oils. Using essential oil of chamomile you can make a chamomile salve or simply buy a chamomile salve and replace all the OTC itch medicines on the market. It is that effective.

Drinking chamomile tea is an effective treatment for hives even. Taking a bath in water that is infused with chamomile essential oil is a great way to get the benefits of this herb over large portions of your body. Steroids are not a great thing to have on you, in you or anywhere near you. Especially children should never be exposed to this potentially life altering drug.

What is Plaintain you might ask. Every year this plant grows in American's yards all over the country and you walk on it and mow it down and it is one of the most effective itch relief plants around. Plantain is a free medicine. Although there are many types of plantain, they are all great to use. English plantain or the broad leaf plantain is most usually used in herbal preparations for itch relief. Once you learn what this plant looks like you won't mistake it for anything else.

You can buy plantain oils and tinctures and salves over the counter and none have the side effects that steroid laden OTC itch creams of the pharmaceutical kind have. Herbal treatments are cheaper and the effects last longer and again, they can get to the actual root of the problem.

What is it? Goldenrod herb

Neosporin/Triple Antibiotic- Goldenrod oil/salve. Tea Tree oil. It's pretty simple folks, mom used Neosporin on us when we were youngsters every time we crashed our bike, when we fell and skinned our knee running across the church parking lot, or after we got into a scuffle with our sibling and ended up with scratches on the face. And that stuff worked! Well it seemed as if it worked miracles, didn't it. Well that was because it allowed no infection to get in which allowed our young bodies to heal themselves and do so pretty darn fast. I am not denying that triple antibiotics and Neosporin works. I will, however, tell you about how much over prescribed antibiotics are in our modern world. The results of this fact are devastating too.

Why do you think we have aids and plagues and staff infections and other "super bugs" that we can not get under control? It has all been linked to the creation. We have essentially created these super bugs, that our modern day antibiotics can not kill, by over prescribing our antibiotics. We put antibiotics in everything! We put it in soap, then we wash our hands multiple times a day, it goes down the drain into our water supply which does not get taken out completely (look it up I am serious) and then we ask why can't our antibiotics kill staff infection?

When you over use a live medication, one that is meant to kill disease, germs, living things, the "thing" is sometimes a virus. Viruses do not die. They multiply, change and adapt. They over come the antibiotics and come back stronger. Now we have a super bug that cannot be killed with our antibiotics. If that happens to you, you die. So you have to think, why not use antibiotics only when it is totally necessary? Problem is, our doctors have, for decades, used antibiotics as a means of preventing disease. Antibiotics are NOT a preventive medication and should have never been used as such.

When we have modern day doctors that are prescribing antibiotics to patients sick with some virus, we have a problem folks! Unfortunately this is the world we live in. So at all costs, <u>except life</u>, stay away from antibiotics. Use herbal remedies to treat sores, cuts, scrapes etc. That could or have become infected. A simple herbal salve made using goldenrod is highly effective at treating the infection and in as little as three days.

You can either buy a salve already made up or you can make your own using goldenrod essential oil or the herb itself steeped for a month or more in a carrier oil, such as olive oil. When it comes to using Tea Tree oil/Melaluca, you just need to buy a bottle from the store, most any store these days carries it, and use it as you would any Neosporin. Tea Tree oil is to only be used externally, but goldenrod can be used internally if prepared for such use, such as an infusion of the dried wild herb. That means it is effective at treating internal infections as well. Just a bit of additional information for you there.

9. Chapter 4: Herbal treatments for children

What is it? My daughter in meditation at age 8

Children are precious. So why in the world has humanity gotten so far away from doing the smart thing, the right thing and turned to drugging children into oblivion? Humanity is so brainwashed today that there are people that actually believe the hogwash that comes out of the mouths of those intent on evil doings when it comes to raising the children of the 20th century. Children are supposed to be hyper and run about, yelling and hollering, playing and carrying on without a worry in their brains! It's called being a child. You were the same. Except hopefully, and most likely since we are talking an older time, your parents didn't look at you playing about and pop a pill in your mouth to calm you. I know mine didn't. People, today, have no idea how to raise a child. It's not because God stopped putting natural instincts into mothers and fathers, it's because they have been brainwashed into believing the garbage that there is something wrong with their child because he or she is hyper and is always full of energy acting like a child all day. You've forgotten how to raise children as a human being! This is true. If it wasn't then explain why there are books and classes you can take and all sorts of informational videos, just to name a few things, that teach you how to be a parent?

What is it? Dead Nettle

The problem is, people have become to dependent on those that say they are in charge. The people have gotten away from self sufficiency and sustainability and can not even function how God created them now. They have been led astray and on purpose. Do you realize the drug industry is a multi-billion dollar a year business, and that includes drugs for children? It's all about the money, and control.... Children should be raised naturally. If there is any human being on the planet that should not be messed with using chemicals, it's our children. At a time when the human brain is at it's all time growth and development they want you to fill their little heads with chemicals and drugs. I say NO!

There are many natural ways you can catch a break from the hectic world your kids can create, with out turning to drugging them into a zombie. I am a parent, I have 4 children that I raised using no drugs. When I needed some sort of response from their little bodies I used natural substances or certain techniques that, as a father, came instinctively to me. If I had a child that did not want to take his or her nap and I felt it was imperative the nap happened, I would simple wait a bit longer so the child could consume an herbal tea that would make them want to take a nap. Naturally calming them down was the key.

If I had a sick child I did not take him or her to a doctor or hospital. I used a natural remedy with 1000's of years of proof that it worked.

Doctors and hospitals have their places in saving lives. But when it comes to simple issues, illnesses and such, these can all be handled at home with God's medicine... meaning herbs and even foods and such things as meditation and other natural healing that uses natural energies.

What is it? Teach your child about natural medicine instead of drugs

Hyperactivity/Anxiety/focus

Although you need to understand that it is quite normal for kids to be hyper, there are times when you want to teach them that it is time to calm down; such as bedtime. Certain herbal teas are usually best for children of toddler age and older. All the herbs you know as calming herbs that you can use for yourself, such as Jasmine, chamomile and passion flower tea, can be used for your children to. That is a very good thing about herbs, you do not have to worry if you can use them with kids or not or what doses to use safely. A simple cup of warm chamomile tea will do wonders for your little one when you want to lessen that natural load of energy.

There are herbal tea companies that make special herb teas for children as well. So finding a favorite tea company and seeing what they have for your little one to help calm them naturally is what you need to do. As a suggestion, a very well known company called Celestial Seasonings is sold in your local grocery stores all over. They carry a wide assortment of herbal teas specifically designed to do certain things for children. One tea called, "*Sleepytime*" is a favorite of mine to be used an hour before bedtime. Then slowly it works to calm the child and see him/her off to sleep naturally. For those stubborn children that need a little something extra and possibly to aid them to sleep a bit better during times when they may be teething or feeling bad from a cold, use *"Sleepytime Extra"*. Sleepytime Extra has Valerian root added in doses that is safe for kids.

Another favorite of mine to calm the little monster is called, *"Tension Tamer"* and it works very good! You can use this to send your child to sleep or during times when you want him/her to be calmer, maybe to watch a cartoon quietly while you catch up with your favorite book. Kids who go to public school can benefit using herbal teas and tinctures that are made to calm and allow more focus. Some of those that can be used to calm and allow more focus is

"Gingko Clarity", by another well known tea company called Yogi Tea and *"Focus"*, by a number of tea companies. If you know your herbs well, you can make your own infusions and save a lot of money. By stocking dry herbs and ingredients used to make your favorite ones you have the ability to make a large amount more than paying $3-$5 per box of about 20 tea bags.

When you are looking for calming, relaxing and focusing herbs here is a short list: Hops, valerian root, chamomile, passion flower, peppermint, skullcap, jasmine, rose hips, basil. Each of these herbs has calming abilities and when someone is more calm their mind calms and they can focus and think better.

What is it? The only health care that will be available after SHTF

Colic/upset stomach/intestinal issues

Babies and younger toddlers sometimes have problems with their digestion system because grown up foods and drinks are just to much stress on a developing system. Although there are prescriptions doctors will give you and over the counter remedies that your neighbor will swear by, let's stick to natural remedies with a proven record of success.

Peppermint will remain my all time favorite for treating any ailment of the stomach for my kids, no matter their age. The only real thing you need to remember, is when you are giving peppermint tea to a child under age 1 just dilute it with some water. Not a 50/50 dilution because you do not want to water down the medicinal ingredients, but maybe about a 70/30 mixture with the higher percentage peppermint tea. Just make up a normal cup of this tea and then dilute that with water before giving it in a bottle. The affects of peppermint tea on colic and upset stomach is simply amazing.

Toddlers love the taste and you can double up on the healing properties of this tea for your older toddler (over age 1) by sweetening it with a bit of honey instead of refined white sugar. When I use this for babies under a year old, I never use any sweetener. Your child will let you know if they approve of the taste or not, although there is nothing wrong with adding a small amount of sugar cane to tea for a younger child.

If you want to try a bit of variety, then any of the digestive herbs can be used. Each of them will practically do the same thing just remember to dilute when necessary and always read up on the herbs you will be using with your family so you do it properly, otherwise using herbal remedies is both effective and safe for anyone of any age. A word about using peppermint as an herbal remedy. Most people do not realize that their beloved peppermint candy is not really peppermint at all, but a knock off of the real peppermint. Now you can get candy made from real peppermint oil and it does taste better and usually the texture of the candy is not as smooth which lets you know you have a candy made using real peppermint instead of the genetically modified (GMO) stuff. Growing up mom never knew the difference. She would hand me a peppermint candy, one of those after dinner mints like you always get at certain restaurants, and say it was for my upset tummy. But it never worked, and mom would say, well it was supposed to work or well it works for me. Problem is, it wasn't the real peppermint, which does work and works very well.

What is it? Usnea. Fights cold and Flu & infections

Flu/bronchitis/pneumonia

How do I know if it is flu, pneumonia or bronchitis, you ask? Well in reality, because you are using herbs to treat and heal these ailments, it does not matter. It does not mater because they all seem to respond to the same treatment no matter which one you are dealing with. Each of these three ailments we are talking about here all have to be treated with antibiotics, usually. Problem is, antibiotics is something that you usually have to get through a doctors prescription. You can only get a doctors prescription if you go to the doctor or the hospital. Now if you are resourceful you might have fish antibiotics at home, which are the same as human antibiotics. But most people won't have these on hand so that is why herbal antibiotic treatment is a necessary practice you should perfect.

There are a few great herbs and herbal concoctions in the world that I would actually endorse if given the chance. They work so well that, to this day, which is going on at least two decades, I have not used any antibiotic other than natural ones. The first one I want to talk about is called Echinacea. Sometimes it's called purple cone flower, because of how it grows. Echinacea is available in a wide assortment of styles. From bulk herb to tea, to capsules, tinctures etc. You name it, echinacea is widely and easily available. It is so popular that it can be found in just about any store anywhere in at least two forms. This should be evident as to it's success rate. People know it works. In fact the very trusted and widely available retail remedy called Airborne is even made using echinacea.

It is safe to use for anyone, anytime whether the person is ill or you are just trying to prevent an illness by strengthening their immune system. The good and interesting thing about all herbs is that they all do more than one job. One of echinacea's jobs is to strengthen the immune system so that a person does not become sick as often.

Then if you are in need of an antibiotic echinacea acts as a decent antibiotic too. So lets talk about how to use it and when to use it as an antibiotic for your children.

What is it? Child herb gardening

Now I like to use an extract of Grapefruit Seed (GSE) paired with the echinacea tea to really power up this antibiotic remedy. I have personally witnessed the power of this combination with my toddlers, children, adults, myself, even my pets! I am talking pneumonia, influenza, bronchitis, upper respiratory infection, blood poison, staph infection, tonsillitis, urinary tract infection and more. So we are not talking a weak deal here. GSE paired with echinacea tea is an infection killer, plain and simple. Echinacea alone, I like to use with a simple cold because it works well and a cold is not something you need antibiotics for. GSE is a very thick, very concentrated extract of the grapefruit seed. Now don't get that confused with grape seed extract, something different all together. GSE is something you must purchase. I can always depend on the internet when I need a new bottle of GSE. Something like $9.97 for a bottle that will last you two years, at least.

This is the proper usage when using it for your young ones.

Infant- Make your little one undiluted echinacea tea and put what he/she will drink in a bottle and drop in just a few drops of the GSE. Do this three times a day for 5-7 days. You should see a change in the child in as little as 3 days. It could take a week to fully heal.

Toddler- Over a year old. Make your child a cup of undiluted echinacea tea and add in 5 drops of GSE. Sweeten with honey and put in the child's drinking container. Monitor the child to make sure the whole amount if drank within the hour, if possible. For the next two doses that first day and the remaining days doses, for up to a week, use 3 drops of GSE per cup of echinacea tea. Drink 3 times a day for 5-7 days.

Child- Over age 5. Make the echinacea tea undiluted with 8 drops of GSE for the first dose. Have the child drink the tea with in the hour. For the remaining 2 doses that day as well as 5-7 additional days use 5 drops of GSE per cup, drink 3 cups a day.

Teen- Over age 12. Have your teen drink echinacea tea three times a day for 5-7 days with GSE. First dose should be 15 drops, with 10 drops each additional cup of tea for up to 7 days.

What is it? Raw honey, natures flu shot.

I always like to use raw honey to sweeten this tea and you will want to use something because the GSE is extremely bitter as it is derived from a citrus fruit seed. So any child over age 1 I would use honey to sweeten. You want to sweeten this tea to the point it takes to consume it comfortably. The GSE does make it hard to drink otherwise. With a child under 1 it is safer to use non GMO refined sugar or raw sugar to sweeten. One finale word about echinacea and grapefruit seed extract to treat infections. NEVER use GSE full strength on your children. Always make sure it is diluted in either a juice of sorts or a tea of sorts because it is highly concentrated and could cause a reaction, not to mention it has to be the most bitter thing I have ever tasted.

What is it? Teach your child to grow a garden to help them focus

ADHD/ADD

A word about this perceived disorder…it does not exist people. In 2016 the man behind this scheme even was recorded, on his death bed, as saying ADHD and ADD are made up disorders that do not exist in order to market to you and teach you to drug your children into a zombie like state. Don't believe me? Look this up on Youtube and you can see the video, in fact there are articles written about this event too. A proper Google search will enlighten you. Again, it does not exist. What you have been told is a disorder, is actually natural childhood behavior. Has humanity gotten that far off track, out in left field to think that a child is not naturally hyper. Hyper activity, lack of focus, short attention spans, these are all natural qualities that most all kids are born with. Until a certain age where they grow up some, this is called being a child.

If you want to calm your child or help him or her focus better, then look into meditation and calming herbs. But don't fall into the brainwashed crowd of failed parents that have turned to drugging their poor kids into unnatural mental states. That is all I have to say about this.

10. Appendix 1

Other works by this author include:

- Living off the grid with Merlyn Seeley – Also available in print
- Money making tips for the writer
- Grapefruit seed extract liquid gold
- An advanced look at Shikata Ryu Ninjutsu
- 12 plants you can not live without
- An introduction to Shikata Ryu Ninjutsu
- Basic meditations for beginners- Also available in print
- Jesus and Buddha two masters one path
- Recipes from the homestead
- Making money with your land

- How to survive being a single parent
- Raising kids in a poisoned world
- Raising chickens for profit
- How to build an Eco-friendly off grid bathhouse
- How to write your first book
- 12 crops that will grow in acidic soil
- A beginners guide to the martial arts
- Ideas for making money on Fiverr.com
- Living off grid a beginners guide
- Letting go of the past a how to guide
- How to simplify your life the ultimate guide
- Living off the grid a beginners guide book 2
- Using herbal remedies
- Planting your organic garden a how to guide
- Homeschooling made easy

- Raising rabbits for profit
- Shikata Ryu Ninjutsu training manual- Also available in print
- Understanding anxiety and depression- Also available in print
- Preparing for the collapse- "Food production basics"- Also available in print

All of these works are e-books that will be easily downloaded onto your NOOK book or Kindle App. Some are available in print and will say that online. Available online wherever books are sold.

11. Reader Notes